Under a Pulsating Sky

Copyright Information

Copyright © 2023 by Anna Kade Schoenbach

All rights reserved. This book or any portion thereof may not be reproduced or used in any manner whatsoever without the express written permission of the publisher except for the use of brief quotations in a book review or scholarly journal.

First Printing:September 2022

Second Printing: February 2023

ISBN: 9798841134473

Imprint: Independently published

Dedication

To my Mother, Father, and Aunt, who constantly cultivate my flowering creativity with their love, even when my mental soil is hostile to the roots.

Introduction

The universe is a strange place, my friends. Our ground is our vantage point, and it is a very limited one. Our ground is far from the only ground that there is, and the gravity that holds us there is far from the only gravity within wells of power. Still, it is from here that we can look up and see a thousand wonders, which are simultaneously at our fingertips and beyond the point where we can ever hope to touch them. For every wonder we witness, there are a hundred simultaneous wonders that we cannot.

There is a rhythm to the universe, not unlike the music of the spheres, a divine melody once thought to keep the planets and celestial bodies turning and twisting in their proper place. These perfect gears have since been shown to be imperfect, as down to Earth as Earth itself. They are made of ice or gas or even oceans of methane, not perfect crystal yet they sing all the same! Just because we can't hear them or see them from beneath our clouds doesn't mean that our suns spiraling, electromagnetic heartbeat doesn't resonate with our planet and send its thrumming out into the great and roaring beyond. It doesn't mean that the quasar isn't there, spinning like a lighthouse, flashing out an incomprehensible message into the endless darkness.

This book is one of a collection of four books of poetry inspired by planets, moons, and other celestial bodies. Each book represents a different theme and can be read separately and in any order.

The science-themed book, "Souvenirs of Ice and Dust," found inspiration from and weaves its lyrical stories around scientific fact right off of the NASA website.

The poems in the introspection-themed book, "Burning Cold in the Quiet Night," are all thoughtful and look inwards into the soul rather than out to the sky.

The mythology-focused "Shattered Stars and Broken Stones" contains poems based on the mythological or literary figures who that have lent their names the moons and the planets.

Those poems that were eccentric, experimental, or just playful are found in "Under a Pulsating Sky," the artistic-themed book. Each poem includes a brief note about the key scientific, mythological, or historical fact that inspired the poem.

We live under a pulsating sky, beautiful in its endless vibrance, inspiring in its sparkling, inky vastness. Those of us who live in cities pollute the sky with light, separating us from the sight of a horizon full of burning meteor fragments, but that light shines outward and makes our planet gleam on the night side with glitters of life. Of a species that forges its own environment and looks out at the universe and reaches out to touch something more.

Is it any wonder that the countless objects, so close at hand yet so far away, drifting in the frictionless grasp of space, inspired me to art? Collected here are poems that drift in dreams, that resist categorization, that travel over weird roads to places far way, where the stars seem red with distance as their light is stretched behind them. Art always has meaning, but let your mind drift like asteroids in the night, or comets on their billion-year journey to immolation, and don't force that meaning to come.

Slow down and relax, and let your mind paint its own picture of eternity.

Table of Contents

Pg 1 - A Universe of Secrets
(Aegaeon, Saturn)

Pg 3 - Lyric
(Comet C/1861 Thatcher)

Pg 5 - Our Father
(Comet 2P Encke)

Pg 7 - Lighthouses
(Neo and Psamanthe, Neptune)

Pg 9 - White Queen
(Triton, Neptune)

Pg 11 - Lady of the Raiment
(Aitne, Jupiter)

Pg 13 - Bloodrite Witch
(Sycorax, Uranus)

Pg 14 - The Umbral One
(Umbriel, Uranus)

Pg 15 - The Oracle
(Ophelia, Uranus)

Pg 16 - The Dark God
(Setebos, Uranus)

Pg 17 - Lady Darkness
(Nix and Hydra, Pluto)

Pg 18 - The Dawn
(Hyperion, Saturn)

Pg 19 - She Who Paid the Ultimate Price (Desdemona and Juliet, Uranus)

Pg 21 - On the Surface (Galatea, Neptune)

Pg 22 - Chang'e (Comet 109P/Swift-Tuttle

Pg 25 - Dragon's Breath (Comet 21P, Giacobini-Zinner)

Pg 27 - The Sarimanok (Cressida, Uranus)

Pg 29 - Protean (Proteus and Hippocamp, Neptune)

Pg 31 - Cordelia, Ella of the Stars (Cordelia, Uranus)

Pg 33 - On the High Sea (Comet Catalina C/2013 US10)

Pg 34 - On the Fault'd Stage (Ariel, Uranus)

Pg 36 - Lagrangian Lovers (Polydeuces, Saturn)

Pg 37 - Husband and Wife (Asteroid 253 Mathilde)

Pg 38 - Falling Away (Comet C/2013 US10)

Pg 39 - A Roaring Voice (Enceladus, Saturn)

Pg 41 - The Fairy King (Oberon, Uranus)

Pg 43 - Queen Mab (Mab, Uranus)

Pg 44 - Trojan Queens (Dione and Helene, Saturn)

Pg 47 - The Search (Dwarf Planet, Makemake)

Pg 49 - Mystery (Planet X, Hypothetical)

Pg 51 - Welcome Party - Comet 1P/Halley)

Saturns' moons have many unique traits - methane geysers, water vapor, and even geological activity. These are all building blocks of life-as-we-know-it, and studying the moons of Saturn could reveal many secrets about our universe and our own home planet.

A Universe of Secrets

Aegaeon of Saturn

Mysteries abound!
There is always more to find.
Just look, you, around
and behold the ruins of your mind.

A universe of secrets,
soft whispers in shivering spaces,
stars in unending bleakness,
lurking in their birthplaces,

waiting to be found.
Perfect spheres in the darkness
mysteries Gaia-bound,
beneath her fertile carcass.

In twilit gloom
and ancient rings,

secrets find their home
and sing.

A song in silence remains unfound.
A universe of imagination - unbound!

Comet C/1861 G1, also named "Thatcher" after the man who discovered it, produces debris that falls through our atmosphere and produces a meteor shower called the lyrids. The meteors seem to come from the constellation Lyra, which was the lyre of the mythological figure Orpheus.

Lyric
Comet C/1861 G1 (Thatcher)

Orpheus sings his solemn song

sparse auroras drip

from taut lyre strings

falling to Earth

through a skein of cloud

it all comes to ash and dust

before it reaches solid Earth

dies in transit

and never even looking back

to distant cold Vega

who never followed his music anyway

too busy with its twin

it makes a poor replacement for his muse

who will never hear his song

who will never see his love

who will never see the sky
yet he sings his lost song anyway.

The nature of art is not to be seen
not to be heard or felt
but made.

Although Comet 2P/Encke is named after the man who discovered it, there is a Sumerian deity named Enki who was the god of water, mischief, crafts, and creation. Stories about him emphasize balance and responsibility and not doing things to excess, and he is portrayed as a wise, life-giving god.

Our Father
Comet 2P/Encke

Nudimmud, our father Enki,

saw our birth.

An attentive father, he visited frequently

to see what we had wrought.

We used the gifts of civilization

to create wonders, and he gave us more,

the falling stars and a bull to guard them,

and rings shimmering,

flaws bright in ice and stone.

We were ungrateful, greedy,

reaching ever further

plundering the jewels set for us

and the gifts beneath

and then turned skyward for our punishment.

But our father is indulgent, smiling,
> "Grow in mind and body,"
>
> "Learn all that there is to know,"
>
> "Eat, drink, love each other, be merry,"
>
> "Create more wonders, my children,"
>
> "And become as many as the stars!"

He loved us so, and, turning inwards,

we found what he saw and brought it out

nurtured it as he did,

now it grows in fertile ground.

Hope, amid devastation, in the air

fed by loving waters:

He smiles.

Very little is known about the moons Neso and Psamanth, though they may have once been the same moon. As potential evidence for this theory, they share the same eccentric, 26-(Earth)year orbit around Neptune. Of the known moons in our solar system, these two are the most distant from their planet.

Lighthouses

Neso and Psamathe of Neptune

Far in the deeps
they wait beyond the
terminating border of sky
and wave

There is no horizon, where
the ferry passes by
carrying you slowly across
the infinite

No island or home
to be seen through
the fog of their
empty vigil

On this distant outpost
looming light flashes over
featureless shifting planes, where
she rests

Mother and daughter's brilliance
pierce the gloom, shining
to reveal treacherous stone
lurking beneath

The stones are knowledge.
The light is knowledge.
The ocean is knowledge.

And you -

You learn.

Triton is a frozen, yet volcanic moon, and is one of the brightest objects in the solar system due to its layer of nitrogen frost, reflecting back 70% of sunlight. It is thought to have once been an object from beyond Pluto, dragged into orbit by Neptune's gravity. It will one day be pulled too close and torn apart, becoming a ring.

White Queen

Triton of Neptune

Neptune brings me down -

Like Pluto, I drifted through space.

Now I twist and turn in the howling winds

of something greater.

Chilling white I prance.

All others take notice: I am the greatest moon!

My wisps of crystalline hair

whip all other moons away.

Cloaked in clouds of chill nitrogen,

bathed in bursting ice and dust,

burning deep beneath,

reaching for the far-off sun's flirtation.

Jealous Neptune draws me in

despite my frosty skin,

as I make eyes at far-off 'molten' Earth

and wink in brilliant ice.

But you moved on to other worlds:

Worlds that might yet have life.

Burgeoning Titan, searing Io, cold Europa...

What do you have that I do not?

Ah, I'll become a ring of Neptune!

Maybe then, torn apart and shimmering,

Earth will realize what they missed!

But by then I'll wink at balmy Venus instead,

and laugh in shivering dust.

Aitne, whose name means "I burn," was a Nymph thought to live under Mount Etna in Sicily, where she aided the god Hephaestus with his forge. The moon takes slightly less than two Earth-years to orbit Jupiter.

Lady of the Raiment
Aitne of Jupiter

Forged from old Carme,

flame striking from her stone,

"I burn, I burn," she roars, and breathes

her life into the forge:

Smelt the earth with volcanic heat,

"I burn, I burn," yet she is cold,

once quenched in ice between stars bright,

twice in dust old and new.

In howling mountain, silent void, and thunder far above:

Two years her forge and hammer shine,

at last in orange swirling clouds –

she lifts her new work high!

A blade to clash the heavens clean,
a rein to hold them fast,
present all things to blazing king,
so that he may rule at last.

Sycorax is the largest moon among the "irregular" moons of Uranus, which orbit in the opposite direction to Uranus' rotation. She is named for a witch mentioned in Shakespeare's Tempest, and, unlike her groupmates Prospero and Setebos, is red in color instead of grey.

Bloodrite Witch

Sycorax of Uranus

Dead before birth,
opponent of mirth,
irregular witch -
her magic marks her red
like blood,
she knows "life's a bitch,"

Two moons grey,
12 million kilometers away.
She was dead before the beginning,
now alive forever – marked in red,
like blood
on her own path, forever.

Sing a song and cast a spell,
The red blood rite, she knows so well.

Umbriel, named for a malevolent spirit in Alexander Pope's 18th century poem, "Rape of the Lock," is the largest moon of Uranus and has a strange, frost-ringed crater. We don't know very much about it.

The Umbral One

Umbriel of Uranus

In the shadow of the shadows,
looming large with gleaming eye,
ringed with frost and
watching.
Seek a ring, a bell, a hope
for freedom from your servitude,
a mystery deep within
bound by chains cruel and deep.
Watch,
oh shadow-borne seeker,
wait until another meets
your ice-cold gaze
and understands
your constrained rage
and pain.

Both planets and moons have gravity. Though the gravity of a moon is small, they can shape the planet they orbit. In the case of "shepherd moons" like Ophelia, their gravity molds planetary rings, preventing particles of ice and rock from floating off into space. They create the light-refracting clusters that give the planets of Saturn and Uranus their distinctive charm.

The Oracle

Ophelia of Uranus

Half doomed prophet // Half shepherd of moons

Half stone // Half ephemeral ice

You whisper in a mirror // standing with crook

To guide others // in stoic silence under

to a quiet sky // an ancient fate,

Watching your charges pass by.

Note: This poem can be read straight through, as just the lines on one side of the dividing marker, or alternating.

When he wrote The Tempest, Shakespeare likely chose "Setebos" as the name for a "patagonian deity" because it was a 'generically exotic name', something that would enhance the "otherness" of the antagonist characters, Prospero and Caliban. Fortunately, the name has, since, been used for better things, such as the scientific name of an Antarctic octopus; Megaleledone setebos.

The Dark God

Setebos of Uranus

Being of a forgotten
(or remembered) age,
a creature of a dark
and unknown place,
for respect,
there is no space
to darken hides
and create a foe,
grey tentacles in icy seas
setting off a perilous glow,
that, to the racist "Other,"
does appease,
and is applauded on the stage.

In Greek mythology, Nyx was the goddess of night, a fitting name for a something so far out in the solar system that the sun is barely more than another bright star in the sky. Hydra, another moon of Pluto, appears to have once been a part of Nyx.

Lady Darkness
Nix and Hydra of Pluto

In the lacuna she lurks with her beast, watching, hungering,

awaiting her feast.

Fragments of a lost, archaic past,

dark Nix will have her time at last.

In nightless night, she stares upon

the gleaming speck of the sun

that one day she claims she will devour.

Yet even the most modest moonlet exceeds her in power:

This fact, though, she will not admit.

Hyperion is named for the father of the sun god Helios and the dawn goddess Eos. Fittingly for such a regal figure, the moon not only remains in an erratic orbit, but is one of the few moons of Saturn that is not tidally locked with the planet.

The Dawn

Hyperion of Saturn

A new day dawns on Saturn,
a shining white
moonlet rises
a mere fragment tumbling forth
traced by a distant star.
Ice mistaken for a sun
shimmering jewel in a saffron sky,
Titan twirls the dancer
this wasp-nest Hyperion,
in a light-footed cluster.
A king free from fate,
a half-comet flying unfettered
ablaze with stolen light
cries out in ceaseless twirling,
to an audience uncaring
in a silent ice field.

Tidally Locked: A phenomenon where gravity forces the moon to synchronize its orbit and always have the same part of its landscape facing the planet

Juliet and Desdemona both died for love - Juliet, stabbing herself in the arms of her dying lover, Romeo, and Desdemona, murdered by her husband under suspicion of her infidelity. Their fate was cruel, and, as moons, they are constantly bombarded by meteors.

She Who Paid the Ultimate Price

Desdemona and Juliet of Uranus

Alone in devastation, together in death -

Oh, how love burns the heart like poison!

Oh, how it stabs like a blade,

crushes like the heaviest frame

of a marital bed betrayed!

Just beyond a god's rings,

twisted - again - by forces beyond control,

pelted by fragments from the beginning of time,

beaten without recourse -

Two women spin in their mutual devastation,

holding each other together

and yet,
they drift
alone.

Highly-cratered Hyperion is very similar in composition to a comet, and does not have enough gravity to pack itself tightly together into a sphere. In fact, the moon is so light in density that it would actually float in liquid water, possibly because of gaps of air or pockets of frozen methane or carbon dioxide.

Galatea is a tiny moon that orbits relatively close to Neptune's equator. Its gravity disrupts the ring system around Neptune, causing "ring arcs" or partial rings.

On the Surface
Galatea, of Neptune

Ripple in the rings
touch-up Neptune as she sings.
Have to look good
as one should,
at all times with Voyagers passing by!

"You have a show in 40 years,
make you the color of tears,
enhance your storms and rings,
(and lock up... other things...)
so that the show may go ever on!"

Paint her canvas bright
stand out in the night
smile as the cameras pass
the illusion made to last
then fade back into the partial rings -
The evidence of her passing.

Chang'e is the name of China's moon exploration program, named after an immortal who was banished to the moon. The reason for her banishment changes depending on the region and the storyteller, but in most versions she is kept company by a rabbit.

Chang'e

Comet 109P/Swift-Tuttle

The rabbit hops onto the moon, grooms himself of silver dust.

"What did I do today?" he asks himself, in the desolate wasteland, "I'm glad you asked!"

"I raced around the sun today,

basked in the blazing light, napped in the sweet, sweet dark,

I was cold and hot, like a snowshoe hare,

as I raced towards the green-blue globe.

I think I gave them a show -

a trailing stream of silver dust fell to the Earth below!"

"And what happened next?"

He thumped his foot on the ground, making no sound in the void,

"A shower of flame streaked their sky,
I watched it from above as I passed by.
Like a fox I was, with a sweeping tail,
better looking, though, by miles.
The people down there watching me,
they oo-ed and ahh-d
and called me the Perseids!"

"I know it's not modest to preen, but,"
The rabbit laughed,
"I know, I am amazing.
I am great. I am fantastic,
and I am oh-so-popular!"
The rabbit looked around him
at the desolate waste,
and his ears began to fall.

"Anyway, I'm tuckered out,
I'm going to go to sleep
for a hundred more years,

so don't wait up for me!"
The comet-tailed rabbit said
to nobody -
nobody at all,
in his burrow on the moon.

The comet Swift-Tuttle is a massive object made of ice, rock, dust, and volatile gasses, which break away as it drifts close to the sun. This debris burns brightly as it falls through the Earth's atmosphere, in this case causing the Perseid meteor showers.

Comet 21P/Giacombi-Zinner has a regular orbit of 6.6 years, and whenever it passes it leaves a trail of debris behind it. When Earth passes through this debris, it causes an event known as a meteor shower, in this case, the Draconids, which appear to come from the constellation Draco.

Dragon's Breath
Comet 21P/Giacobini-Zinner

The little hatchling dragon of the north

puffing shimmering smoke-rings in the night,

his mother's regal arrogance feeds her pride

for her hatchling son.

Small Draconid's breath was weak and faint, yet,

some nights he was strong, his breath a torrent,

a blazing silvery storm of starfire

of cold-burning ice shards.

"Halloween is coming, my little one!"

"Harken to the North, where cold cruel winds rise."

"You were born for this: Beat your wings, breathe deep!"

"Hasten Winter's breath!"

Small Draconid tried, blowing fanciful snowflakes.

Not enough for mother, who gave forth a gale,

her voice a rampaging Autumn storm, she said:

"You are not my son."

No expectations as he left his nest,

the hatchling was free to faintly breathe at last,

harmless storms of meteors, burning cold

in the quiet night.

The Sarimanok is a legendary bird from Filipino folklore that is considered a symbol of good fortune. It may also be the inspiration for Nintendo's 4th-generation legendary Pokemon called Cresselia, a warrior that upholds the natural order, with many features inspired by the crescent moon. The name Cressida, indeed, comes from "crescent" and from "Troilus and Cressida," a play by Shakespeare about two lovers during the Trojan War.

The Sarimanok

Cressida of Uranus

Here at the edge of the universe
there be monsters:
the "Ugly Duckling," dull-plumed,
rock and dust and ice
becoming a gleaming swan
with crescent wings.

Cygnus sibling, warrior moon,
a crest, a feathered tail,
kaleidoscopic wings
slicing the sky
with otherworldly fortune.

Sarimanok's call wakens us to war -
the war of life itself!

The she-monster,
gliding on auroric wings,
hands us her lunar swords
and bids us fight on
for a hope as yet unknown.

Hippocamp, named for the hippocampi, the mythical water-horses of the Greek god Poseidon, is very close to Proteus. Though Hippocamp orbits faster (23 hours to orbit Neptune, instead of 27) it might actually be a fragment of Proteus, broken off in an ancient collision.

The word, "protean," means "changable" and, if this dark-colored moon Proteus had just a little more mass, it could morph into a sphere. However, it remains a lumpy, scarred moon.

Protean

Proteus and Hippocamp of Neptune

Shapechanging dark horse

gallops through the stars

become a box,

become a serpent, become a sphere

- if you can.

Expanding mouth like a gulper eel

to eat the moon, to eat the stars, to eat the sun,

eat the azure gas of Neptune.

Never to be found again in the same form

in the same way,

but the pace is predictable

and repeats every long day.

Largest and smallest,

and peaceful and fierce,

comet-chip melting into new shapes

and forms

to make a horse-tail aurora that swats at the flies

of dust and stone

that never became rings,

that never glitter in the hard-lit sun

in the lost depths of a gas ocean

  – ever more.

Cordelia is one of Uranus' "shepherd moons," which keep its rings in order. She is named after the youngest daughter of King Lear in the play of the same name, who, despite her father's petty cruelty, remains loyal and loving to him.

Cordelia, Ella of the Stars

Cordelia of Uranus

Have you heard the music of the spheres?

Passing by the room where you wait,

dressed in your modest dress of forgiveness

as all the suitors, and your father too, pass you by.

Have you heard the music of the spheres?

Alluring and oh-so-distant,

because you spoke the truth of love,

not base flattery dressed as love,

and wished only for happiness.

Have you heard the music of the spheres?

As the moon-mad and sun-blinded man

staggers to you and falls at your feet,

and you lift him up with new-returned love,

and you dance at last, caring not what he *deserves*,

because your kindness is as vast as the heavens above.

You dance to the music of the spheres,

guiding his every step as you carefully shepherd

great Uranus' rings into an order,

away from notice,

but no less important, no less loved

than great ladies

in star-sparked celestial cloaks,

who dance on the ballroom floor without ever knowing

that they are guided by your leading hand.

Carbon-rich comets like Catalina C/2013 US10 might have been an important source of carbon during the formation of Earth. The comet was "slingshotted" around the sun in 2015 and is now moving fast enough to eventually completely escape the solar system.

On the High Sea
Comet Catalina C/2013 US10

A black boat full of shifting life

sleeping in its ancient form

watching for a buyer

finding only glut.

The captain shrugs and soldiers on

speeding towards the outer edge

past the sun's watchful eye

to new trading posts ahead.

On the Fault'd Stage
Ariel of Uranus

Young, yet creased and crevase'd,
age'd beyond her eternal years
breathing star-heat in and out
as she is pushed and pulled
in her most-obedient track.

Aeriel shimmers in her depths with
strange dramas on her surface -
chronicles of lightness and density
and eons of unrelenting service,

to the blue one,
to the wizard,
to nature's force,
to light and dark.

With youthful voice and wizened hand
she spins her tale to her audience
seated high in the furthest seat
and closest to the blazing spotlight.

Ariel is a character in the Tempest, and the moon has a few unusual properties: It heats and cools quickly, has a porous surface that reflects light oddly, and may have been heated enough to have a "core" of heavy matter.

Celestial bodies have their own gravity, and they pull on each other. Lagrange points are locations in space where this gravitational tug-of-war is equal, and a third object - like a spaceship or asteroid - can remain in place without any outside force.

Lagrangian Lovers

Polydeuces of Saturn

Gravity is love's pull -

A love shared
with trojan heart
two points in balance
one moon and another
and another and
a giant great beyond
with wandering heart
co-orbital lover orbiting
another trailing another
in perfect balance
stable -

a multi-heartbeat love.

Asteroid 253 Mathilde, like many celestial objects, is named by its discoverer, astronomer Moritz Loewy, the Vice Director of the Paris Observatory. In this case, it is named after his wife.

Husband and Wife
Asteroid 253 Mathilde

A lazy day on Mathilde

is more than half a month -

Battered a bit, not broken,

though her husband is so married to his work

that he never comes home

eyes fixed on the stars

that now bear her name

as if that will absolve him!

Polydeuces is locked in a Lagrange point, where the gravitational pull between the planet - Saturn - and a larger moon - Dione - is perfectly balanced. The moons Helene, Mimas, and Enceladus also interact, in various ways, with Dione and Polydeuces through their shared gravitational pull.

A few million years ago, Comet C/2013 A1 Siding Spring was dislodged from the vast Oort cloud in the far outskirts of the solar system and began to fall towards the Sun. On October 19, 2014, it just barely missed colliding head-on with Mars. It has since left the inner solar system, and will not be back for 740,000 years.

Falling Away
Comet C/2013 A1 Siding Spring

The angel from the heaven beyond

did not fall to Earth,

but was nudged from its complacent throne

to a far redder place.

Iron in the sand called to their icy wings

which beat laboriously before

they could crash to ground,

and fearful Phobos gasped

in the feathers left behind

that burned briefly in Mars' thin skein

as the master's grace left it far behind

and fell to a new desolation.

Enceladus is a brilliant white, extremely cold moon that has a vast ocean. Geysers of water (along with carbon dioxide, methane, ammonia, and nitrogen gas,) erupt out from under the ice, powered by warm fissures in the crust at the bottom of Enceladus' ocean. These geysers create a mist of ice that trails behind Enceladus, building up one of the rings of Saturn.

A Roaring Voice
Enceladus of Saturn

The ice lion roars
chill breath blizzard accusing,
"How dare you!"

"Say no more about your good motives,
still your words,
you privileged fool!"

Brilliant blue eyes
glaring at my lies
judging them, freezing them.

"Good words are meaningless,
because you get to leave,
back to your happy life!"
Ice mane shimmering

with an aura of truth,
"For everyone else, it never ends,"

His snarls haunt me
but the ice lion is trapped in resonance,
his roars pulsing, ineffectual

a ring of ice born of his blood
shimmering in the darkness -
"I can be more," I say.

My words freeze into useless dust:
but he can only howl
and I can cover my ears and move on.

In the play "a Midsummer Night's Dream," Oberon is the fairy king who is put to sleep as a part of the trickster, Puck's great prank. The moon, Oberon, is heavily cratered but also has a huge, 6km high mountain on its surface.

The Fairy King
Oberon of Uranus

Ah - hear the pockmarked fairy king!
he blows his hunting horn
and stalks across the sky
in his wild hunt.

Ah - how he brags!
"I shall bag the greatest stag!"
But no stags lurk in Uranus' clouds,
where only Voyagers dare roam.

Ah - see the trickster smile
as skein after skein of drugged wine
make the king teeter on his mount
and fall before the great mountain
in a pile of stellar snores.

Ah - See who truly rules,

for when the king sleeps,
the fairy nation prospers!
And things get done,
deep in the cosmic mist -

"For those who rule,"
the fey trickster says,
"are always the truest fools,
and 'tis our job to keep them happy,
and do what needs to be done."

Mab is small and dark in color, so it blends into most images taken of Uranus and its moons. Voyager 2 completely missed it during its flyby in 1986! Named after the Queen of the fairies in English folklore, Mab was only caught by the Hubble telescope in 2003.

Queen Mab

Mab of Uranus

The abandoned Queen of Rings,

of diamonds (not frozen hearts,)

overlooked until,

on ground also overlooked,

she was seen by a far-off viewer.

Yearning, they cannot meet

in her sterile queendom of dust

and shimmering rings

or on his *terra firma*

burgeoning with life -

So they pine.

The moon Dione is in resonance with the moons Enceladus and Mimas, which means that its gravitational pull causes the moons to speed up or slow down, keeping them in the same positions relative to each other. Dione is constantly bombarded by fine ice powder from the E-ring of Saturn and encircled by huge canyons that gleam white, possibly due to water ice.

Trojan Queens

Dione and Helene of Saturn

Why do you hold me back?

Because she is a queen,

oh princess,

and you would bombard her,

steal her crown

and lose it in the endless garden of glittering rings.

But I create the rings! It is my garden!

So it is, and you may do what you wish in it,

but oh princess, let her rest

her shattered face

in resonance,
in this stability
let her rest!

That is no fault of mine!
We both hurt
both sharing an orbit
sharing our cold fates
both bombarded constantly...
Why don't you understand?
Why do you hold me back?

Princess, let her have a moment,
because the best parts of the odyssey
are taken by the hero
alone.

The best parts of my journey have been with my mother!
The best parts of my journey have been with my queen!
You don't understand, you don't

understand:
We wake each other from our slumbers,
steal our icy crowns
and race, laughing,
across our garden amidst the stars.

Helene is named after Helen of Troy, a half-divine daughter of Zeus said to be so beautiful that she inadvertently was the cause of the Trojan War. She shares an orbit with Dione, which is made stable by the gravitational pull between Saturn and Dione.

The dwarf planets in the solar system often have unique properties, and Makemake is no exception - it has a surface comprised of snow-like pellets of frozen methane or ethane covering its surface. Scientists want to study many of these small worlds for their life-developing potential, though no extraterrestrial life has been found yet. Interestingly, most of these dwarf planets are named after life-giving fertility gods.

The Search
Dwarf planet, Makemake

Worlds of snowy ice
deserted tundras
tiny and fragile
that water would boil -
what do I see but

Life?

Lost amid the blue planet's seas,
in realms of ice and stone cold
for the traveler across eons,
ever cast away from the sun's glow -

    hiding, surviving, thriving...
    hiding, burrowed in rock.

hiding, smothered in oceans.

in the vents of volcanos.

in the sulfurous streams.

in the strangest of places...

Life will always find a way.

Intelligence, though less certain, will seek in the

far off reaches of the universe,

ending in the places closest to itself, in

Life.

Why else would we give

names of fertility to these lonely places

if we had no hope

for their beating hearts

to match ours in kind?

Planet X is an unobserved Neptune-sized planet theorized to orbit the sun beyond Pluto. Planet X's gravity might explain why some objects in the Kuiper belt sometimes behave oddly and cluster together, but it also might not exist.

Mystery

Planet X, Hypothetical

You, oh mystery

in depths where light fails and gravity reigns supreme.

Oh, master of obscura,

sleeping beyond the cosmic debris

of the great wreckage,

the shadow of spawned sterility and life,

numinous and finite in mortal reckoning.

A star voyage, deepened by lingering drifts,

that never saw the tiny mortals

on a tiny blue planet.

If you even exist in your place

beyond the underworld's gates,

sleeping in your Ry'leh,

with dreaming gates open to clusters of gossiping worlds,

those whispering murders of crows,

who then wake and turn to the little blue world,

watch with hungry eyes,

and caw one word:

"X"

Halley's comet has a period of 76 years and is recorded throughout human history. When it last passed by Earth in 1986, it was joined by an international space fleet (including spacecraft from Japan, Europe, the Soviet Union, and the US) that studied the comet, took samples, and gathered data.

Welcome Party

Comet 1P/Halley

We have assembled here today to greet you -

We have watched you from far below and tracked your shining light -

We have used you as a metric, as a ruler, for your orbits of night -

And we are throwing you a party

because you mean so much to us.

Celebrate with us all for a while

with Susei and Sakigake and Vega 1 and 2,

ISEE-3, Giotto, Pioneer 7 and 12,

become drunk on the attention of a whole planet united

to revel in you, collect souvenirs of old ice and dust.

Tell us, are there more messages from the stars

for us?

You gleamed bright over the Battle of Hastings,

curling your tail over many omens great and small,

once, they thought, you would pass through and leave us forever,

but Halley predicted your return,

and 16 years after his death,

you came back and then back again.

To your next passing, old friend,

to when your eternal face turns towards us again.

In 2062, where will we be, when we next meet?

And when you go,

let's raise a glass

and celebrate together again!

Acknowledgements

I would like to thank NASA for unravelling some of the mysteries of our observable universe, and for hosting a great site. Without them, I would be unaware of most of the subjects of my poems.

My mother and father again deserve my gratitude for their never-ending love and for their feedback at various stages of this book. I would also like to thank my high school English teacher at the Nora School, Christopher Conlon, who is the reason I am a writer in the first place. Without him, I would not be anywhere close to where I am now.

I would like to acknowledge all of the art teachers I have had over the years. Though I have forgotten most of your names, each one of you has taught me the meaning of creativity – either because you nurtured it, stifled it, or showed me a new way to use the mediums available to me. The projects that came out of these classes have found their way back into my poems: from the misshapen paper mâché gargoyle that hung in the entrance hall of my middle school during my time there, to the charcoal and graphite panorama of dancing fairies that grace my childhood bedroom, to the small clay vase made to look like a tree stump blossoming with mushrooms… all of these are a part of me, and a part of my work. And I thank you for it.

I also thank my reader and wish them luck in their own creations. We are all capable of art, no matter what we might think of ourselves.

Author Bio

Anna K. Schoenbach is a freelance science and medical writer and editor who occasionally takes a break from writing articles about science to try to explore science through poetry. There is a wonder to science and even medicine that is often forgotten, and direly underappreciated.

She lives alone in Maryland with her two cats, and often finds herself looking out of her window and wondering about the lives of people below. A classic introvert on the autism spectrum, the stars above, barely seen above the tree- and building-lined horizon, are almost easier to understand.

www.ingramcontent.com/pod-product-compliance
Lightning Source LLC
Chambersburg PA
CBHW070317220526
45465CB00004B/1889